Spring Meeting Poster Session and Networking Reception 2022

Held at the 2022 AIChE Spring Meeting and 18th Global Congress on Process Safety

San Antonio, Texas, USA
10 – 14 April 2022

ISBN: 978-1-7138-6833-0

Spring Meeting Poster Session and Networking Reception 2022

Held at the 2022 AIChE Spring Meeting and 18th Global Congress on Process Safety

San Antonio, Texas, USA
10 – 14 April 2022

AIChE
The Global Home of Chemical Engineers

Printed from e-media with permission by:

Curran Associates, Inc.
57 Morehouse Lane
Red Hook, NY 12571

AIChE
120 Wall Street, FL 23
New York, NY 10005-4020

Phone: (800) 242-4363
Fax: (203) 775-5177

www.aiche.org

Additional copies of this publication are available from:

Curran Associates, Inc.
57 Morehouse Lane
Red Hook, NY 12571 USA
Phone: 845-758-0400
Fax: 845-758-2633
Email: curran@proceedings.com
Web: www.proceedings.com

TABLE OF CONTENTS

640206 Hydrogen Plasma Reduction of REOs/Salt for Rems Production

Abstract:

Rare earth elements (REEs) are crucial to the production of green energy motors in wind turbines, hybrid cars, electric vehicles, and electronic devices. REEs are also used in defense systems, including precision-guided munitions, radar systems, avionics, night vision equipment, satellites and guided missiles. China dominates the REEs production and exports and is continuing to reduce REEs exports. DoD and DOE are looking at options to increase domestic production in the U.S. However, rebuilding the U.S. rare earth industry could take many years.

In this paper, we will report a process for producing REMs from REOs by plasma-enhanced hydrogen reduction to recycle rare earth materials from post-consumer products. Of the five critical REEs, [dysprosium (Dy,66), neodymium (Nd,60), terbium (Tb,65), europium (Eu,63), and yttrium (Y,39)]. This is an efficient and low energy consumption technology for five most valuable REMs (Nd, Y, Dy, Sm, Ce) production from their corresponding oxides.

639549 Study of Surfactant/Nanoparticles Formulations in Rock/Fluid Interaction for Enhanced Recovery of a Colombian Heavy Crude Oil

Abstract:

Heavy crude oils have been considered as an alternative that could offset the world's energy demand and have been widely studied in recent years. However, this type of crude oil is characterized by being asphaltic, dense, and viscous, with a low API gravity (10 to 22 °API) and high viscosity (>100 cP) in reservoir conditions, as well as its hydrocarbon fractions, tend to adhere to the surface of the rock, causing low mobility compared to conventional crude oil, which generates difficulties in its extraction and transportation processes, making the recovery factor of this type of crude oil low. For this reason, Colombia, occupying fifth place in possession of heavy crude oil reserves in Latin America, has focused on the study of enhanced oil recovery (EOR) techniques to increase the yield of these crude oils in oilfields in a more economically accessible way.

Currently, hybrid techniques such as injection of surfactant/nanoparticle formulations are being studied due to the properties of each of their components. In the case of surfactants, these are used to increase the viscosity of the injection water, making the mobility ratio between water and crude oil increase, but unfortunately, they tend to adsorb to the rock generating a low efficiency in crude oil recovery. To ameliorate this problem, it has been found that dispersions with nanoparticles (NPs) can modify the wettability of the porous medium and transport the surfactant to the oil/water interface instead of being adsorbed to the rock, making the oil encapsulated in the rock more easily released by forming highly stable emulsion with low viscosity and oil/water interfacial tension.

In this study, the effect of formulations composed of a surfactant *S* with silica (SiO_2-NPs) or alumina nanoparticles (Al_2O_3-NPs) on the recovery of a Colombian heavy crude oil of 16 °API @ 25 °C in a sandstone-type rock was evaluated. For this purpose, detergency tests were performed at 60 °C for 120 min, where solutions containing 10,000 ppm NaCl and 5,000 ppm surfactant S were used in the presence of NPs with concentrations in the range of 0 to 1,200 ppm. Finally, contact angle tests were performed to determine the change in rock wettability after the use of the formulations.

The results showed that the recovery percentages of the crude oils in the presence of the surfactant were in the order of 43 - 77 % w/w, while the use of the surfactant formulations with NPs promoted recovery yields in the range of 63 - 97 % w/w. Similarly, it was found that the rocks changed their wettability from oil-wet to water-wet after interacting with the formulations. Finally, the effect of the SARA fractions of the crude oil used on the mineral surface was studied, determining that resins and asphaltenes present a greater interaction with the porous medium, but with the application of this technological innovation rock/fluid interaction is reduced, suggesting that surfactant/NPs formulations can enhance crude oil recovery with a high potential for scaling up.

Based on the above, it has been proven that nanotechnology has entered the oil industry providing multiple benefits such as altering the rock wettability and reducing the oil/water interfacial tension, which allows increasing the recovery percentage of heavy crude oils, thus understanding its importance in EOR processes.

639553 Analysis of Catalytic Aquathermolysis on the Physicochemical Properties and Maltenes-Asphaltenes in Colombian Heavy Crude Oils By Using 1H-NMR

Abstract:

Catalytic aquathermolysis is an enhanced oil recovery (EOR) technique that uses steam injection with a catalyst to produce crude oils in reservoirs with lower energy requirements and less time than conventional methods. Therefore, the crude oils obtained are of better quality due to the new physicochemical properties and composition of maltenes (saturates, aromatics, resins) and asphaltenes. In this work, the catalytic aquathermolysis process was studied in two Colombian heavy crudes. The reactivity tests were carried out in a batch microreactor at 270°C and saturation pressure of 800 psi, for 66 hours, using iron naphthenate at 100 ppm with respect to the metal ion. The use of the catalytic aquathermolysis technique reduced the gas yield in the range of 1 to 3 % w/w, the viscosity from 35 to 60 %. Additionally, distillation curves (ASTM D7169) of the improved crudes showed conversions of the 525°C+ heavy fractions between 4 and 8 %, showing a significant decrease in complex fractions with high boiling points.

Analysis of the average molecular parameters (AMPs) via 1H-NMR spectroscopy confirms that the reactivity and products yield vary according to the complexity of the molecular structure of the maltenes and asphaltenes present in the heavy crude oils. The average molecular parameters showed that the major differences concerning the base crude were aromaticity and alkyl chain length. During the application of this recovery method, it was established that the selectivity and reactivity depends on the composition of the heavy crude oil. Finally, the improvement of the physicochemical properties of crude oil at reservoir conditions is a complementary recovery mechanism to the conventional steam injection process, and in turn, allows the transformation of complex fractions towards the formation of species of smaller size and molecular weight.

639575 Application of Liquid Products from Polystyrene Conversion As Pour Point Depressants in a Colombian Paraffinic Crude Oil

Abstract:

Saturated hydrocarbons known for their difficult treatment due to the content of paraffin composed of long carbon chains from C18 to C60 require a complex study where properties such as crystallization point, viscosity and pour point are analyzed; these are important when studying the nature of the crude oil and must be controlled from the reservoir to the surface. In the absence of an adequate study, these molecules are often deposited when the temperature at which they are exposed is equal to or lower than their crystallization temperature, causing additional costs and large-scale operational problems. In response to this problem, the use of scrapers or heaters has been useful to avoid the agglomeration of paraffin, but these conventional techniques are not completely cost-effective. Therefore, research on recent technologies on the use of pour point depressants (PPD) is gaining importance, as they improve the flow of crude oil easily and effectively by lowering its pour point, preventing the formation of paraffin deposits. The proposed alternative considers the degradation of plastic products by the pyrolysis method and its use as a pour point depressant in paraffinic crude oils, for which polystyrene was disintegrated at 425° C and based on pour point and viscosity tests, its effectiveness was evaluated at different concentrations in two crudes from the Colorado Field, known for its latent problem of precipitation and formation of paraffin deposits. The liquid products obtained from the pyrolysis of the base crude were able to lower the pour point of the base crude by 3 °C with a minimum concentration of 0.5% by volume. The results determined that toluene and the liquid products obtained from the pyrolysis of polystyrene could be postulated as depressants, provided that the corresponding analysis of their application is carried out and the conditions for their use are studied in greater depth

638929 Effect of Initial Water Saturation on Carbonate Rock Wettability: A Contact Angle Study

Abstract:

Carbonate reservoirs hold more than half of the world's remaining oil. Wettability has been proven to have a significant influence on oil recovery from carbonate reservoirs. In most cases, carbonate rock tends to be intermediate-wet or oil-wet due to positive surface charge that induces the adsorption of organic materials. Besides factors including oil composition, temperature, rock mineral composition and so on, the resulting wettability of rock is also affected by initial water saturation. The main objective of this work is to find out how initial water saturation affects carbonate rock wettability.

Indiana limestone core plugs were cut into slices for the purpose of contact angle measurement. Filtered crude oil from the field was used for the oil-aging process. An oven was used to raise oil temperature to accelerate the oil-aging process. A vacuum pump and a desiccator were used to saturate the slices with water or oil. A pressure-plate apparatus was used to establish different initial water saturations. Ambient contact angle was measured using the device Attension Tensiometer.

Two similar curves describing the relation between initial water saturation and the applied air pressure were obtained. Both show a rapid saturation reduction from 100% to 80% in the beginning and a minor reduction as air pressure went above 40 psi. The lowest water saturation value was around 41%, while the air pressure was 66 psi. Different initial water saturation values resulted in different wettability, especially when the oil-aging process was limited to a few days. Generally, samples with lower initial water saturations had a stronger tendency to become oil-wet. After 12 hours of oil-aging, samples whose initial water saturation was lower than 60% became strongly oil-wet. Samples whose initial water saturations were between 60% and 80% became oil-wet. However, samples fully saturated took more than 4 days to become oil-wet. For samples that had high water saturation values, the time required to obtain strong oil-wetness was longer. Samples' oil-wetness was found to be positively correlated to oil-aging time.

This study shows the effect of initial water saturation and oil-aging time on carbonate rock wettability. It also provides a method to have some control on rock slice wettability. This method can be helpful in studies related to wettability.

638931 Influence of Permeability on the Relative Contribution of Wettability Alteration and IFT Reduction in Oil Recovery

Abstract:

Carbonate reservoirs contain a significant fraction of the world's remaining oil. However, carbonate rock tends to be oil-wet as a result of long-time contact with oil in most cases. As a result, wettability alteration is found to be the other major EOR mechanism besides IFT reduction. Some studies have shown that the contribution of wettability alteration is affected by rock permeability. The main objective of this work is to clarify how the rock permeability affects the relative oil recovery contribution of wettability alteration and IFT reduction in carbonate rock. Indiana limestone cores of three different permeability ranges were cut into plugs and slices. They were aged in filtered crude oil under high temperature in the oven for two weeks. A locally synthesized cationic gemini surfactant was used to alter rock wettability towards the water-wet condition. Both ambient contact angle measurement and spontaneous imbibition experiments were conducted for comparison. Results showed that permeability is negatively correlated with the oil recovery contribution of wettability alteration. In spontaneous imbibition tests, lower permeability samples produced a larger fraction of oil in place. This was because capillary pressure helped the imbibition of water phase when rock wettability became water-wet. Lower permeability is related to a smaller pore throat radius and thus implies larger capillary pressure. As a result, in lower permeability rock, the relative oil recovery contribution of wettability alteration was enhanced compared to that of IFT reduction. In the ambient contact angle test, however, slices of various permeability showed comparable contact angles before and after wettability. This observation can be explained by the negligible effect of capillary pressure under such cases. This study shows that when rock wettability is altered from oil-wet to water-wet condition, the contribution of wettability alteration on oil recovery rate is negatively correlated with rock permeability. Besides, this study also suggests the locally synthesized cationic gemini surfactant as an effective wettability modifier. Conclusions obtained in this study indicate that more attention should be given to wettability alteration when optimizing the EOR fluid formula for low permeability carbonate formations.

638932 Development of Magnetic Surfactants for Oilfield Applications

Abstract:

Commercially available surfactants, including polyethylene oxide ether, aryl sulfonate, and petroleum sulfonate, exhibit stability issues when applying high salinity (120,000-220,000 ppm) and high temperature (≥90°C) carbonate reservoir. The severe reservoir environment causes surfactant decomposition and decreases their performance in lowering interfacial tension (IFT), altering rocks' wettability, and mobilization.

For addressing surfactant stability issues, magnetic surfactants were synthesized. The chemical structures were confirmed with the aid of NMR (^{1}H, ^{13}C), FTIR, and mass analysis. The heat stability and aqueous solubility tests were assessed using Thermal Gravimetric Analysis (TGA) and aging methods. The spinning drop method was used to identify interfacial tension. The surface tension and critical micelle concentration (cmc) were measured with a force tensiometer (Biolin Scientific). The rheology was performed on the discovery hybrid rheometer (DHR-3).

The surfactants exhibited good solubility, and the TGA graph showed the decomposition temperature of the prepared surfactants above 230 °C which is superior to oilfield temperature (≥90°C). Similarly, the aging technique revealed no change in the chemical structure after putting them in the oven at 90 °C for one month. The cmc, surface tension at cmc, and interfacial tension of the surfactants were comparable to industrially applied surfactants. According to rheological data, the storage modulus was decreased by enhancing the surfactant concentration at a lower rate of shear and frequency due to charge screening and polymer interaction.

The developed magnetic surfactants exhibited remarkable aqueous and heat stabilities and demonstrated immense potential in severe carbonate rocks.

639300 New Crude Oil Viscosity Models for Burgan Oilfield

Abstract:

In the petroleum industry, the prediction of viscosity using correlations is commonly used for solving many reservoir engineering and surface production problems. In the last seven decades, several black oil viscosity models have been developed. However, these models may only be suitable for the geological regions for which they were developed and may not provide acceptable results for crude oils for other regions.

The purpose of this paper is to create a comprehensive oil viscosity models for the Greater Burgan Oilfield-Kuwait which is the second-largest oil field in the world with estimated recoverable oil of nearly 6% of the world's proven conventional oil reserve. Therefore, 824 data points of oil viscosity measurements for the Greater Burgan Oil Field has been used to evaluate the accuracy of the published models and later used to develop a new oil viscosity models for estimating dead, saturated, bubble point, and under saturated oil viscosities. The validity and accuracy of the proposed models were confirmed by comparing their predictions with 166 measured data points from the field itself.

The new models will improve our predications of oil flow rate, pressure drop in tubing and flow lines and hence the design of production equipment and surface facilities. These models can easily be integrated into reservoir simulation for planning primary, secondary and EOR studies.

640000 CO_2 Dissociation over Ni Supported on Doped Ceria Catalysts

Abstract:

CO_2 dissociation to CO is an interesting fundamental reaction that can be part of the chemical looping process which utilizes natural gas and CO_2/H_2O to generate synthesis gas with solar power. Based on previous research, Ni-$CeZrO_2$ synthesized via various methods, including the combustion methods using citric acid, is a promising material for CO_2/H_2O reforming of CH_4. This study focuses on using temperature programmed reduction followed by a pulse CO_2 dissociation reaction to characterize and evaluate the synthesized catalysts. The reduction characteristics and pulse reaction activity and stability of catalysts are summarized and compared. There are optimum Ni and Zr loadings for the activity, capacity, and stability of Ni-$CeZrO_2$ for CO_2 dissociation. Various pre-treatment methods, such as non-thermal plasma treatments, are studied to modify those catalysts for improved CO_2 dissociation.

640134 Efficient Degradation of Naphthalene Using Hydrazine Assisted Nano-Iron Peracetic Acid Process

Abstract:

To improve the oxidation capability, a well-known reducing agent, hydrazine (HZ) was introduced in the nano-iron/peracetic acid process (NI-PAA). Significant oxidation performance was observed in the HZ-NI-PAA process for the degradation of naphthalene (NAP) in a wider range of pH values (3-9) even with the low dosage of NI i.e., 0.5 μ M. The reactive oxygen species (ROSs) were determined to quantify the presence of hydroxyl radicals (OH) by using electron spin resonance (ESR) technique. Benzoic acid (BA) was used as probe indicator confirming the higher intensity of hydroxyl radicals in the NI-PAA system for the naphthalene degradation. The conversion of Fe (III) to Fe (II) was significantly improved by the facile addition of hydrazine. The decomposition products of hydrazine include nitrogen gas, nitrogen dioxide and nitrous oxide whereas, nitrogen gas was the major environmentally friendly decomposition product among all. Intermediatory products were also determined that proposed three possible different pathways for the degradation of naphthalene. Higher dosages of NI and PAA significantly increased the naphthalene degradation. The humic acid showed inhibitory effect on the degradation of naphthalene whereas, anions (HCO_3^-, Cl^-, SO_4^{2-}) have minor effect. Natural water was used to evaluate the oxidative capacity of Hz-NI-PAA process and further investigations were also performed for the degradation of different micropollutants to find its application. This study significantly improved the NI-PAA process and its possible application in water treatment.

640136 Copper Oxide and Sodium Percarbonate System Supported with Iron Nanoparticles for the Simultaneous Degradation of Trichloroethane and Hexavalent Chromium

Abstract:

Copper oxide and sodium percarbonate system supported by iron nanoparticles was investigated for the simultaneous removal of trichloroethane and hexavalent chromium in the wastewater. The degradation efficiency was evaluated by finding the initial effect concentrations of chromium, trichloroethane and sodium percarbonate as well as the time intervals of sodium percarbonate addition to the aqueous solution. Experimental results confirmed that the removal rates of chromium and trichloroethane reached significantly up to 88.9 and 92.30 % respectively with an optimum dosage of 0.5 mM copper oxide and 1.5 mM Sodium percarbonate. It is presumed that copper oxide and sodium percarbonate promoted the reduction of hexavalent chromium and oxidation of trichloroethane that was completely controlled by iron-nanoparticles. The enhanced degradation of hexavalent chromium is attributed to the availability of the reactive sites of nanoparticles and production of ferrous ions increased the aqueous solution by dissolution and corrosion of nanoparticles. The synergetic role of copper oxide/sodium percarbonate/ iron nanoparticles was observed and identified as this system behaved as an activator for sodium percarbonate as well as reductant for the hexavalent chromium for their degradation purpose. Finally, this system has a great potential for the remediation of chlorinated organic solvents and heavy metals from the wastewater streams.

638294 CO_2 Capture and Conversion to Methanol over a Copper Based Dual Functional Material

Abstract:

The industrial intensification has significantly increased the greenhouse gas emissions, among them carbon dioxide, which is responsible for severe environmental issues, such as global warming and climate change. In this study, a novel bimetallic dual functional material based on 10%Cu/10%CaO/Al_2O_3 was developed to capture and convert CO_2 to methanol. Materials with fixed composition was synthesized in the laboratory and characterized using XRD, XPS, BET, and TGA. CO_2 capture experiments were performed using a magnetic suspension balance up to 20 bars. Meanwhile, conversion tests were performed in a high pressure packed bed reactor at 300 °C and 40 bars. The obtained results in term of adsorption capacity revealed a high CO_2 uptake. While CO_2 conversion tests showed a higher selectivity towards methanol and a good conversion rate.

639224 Research on the Influence of Wax, Resins and Asphaltenes on Wall Sticking of Shale Oil

Abstract:

There is an urgent need to increase the development of unconventional oil resources such as shale oil to ensure the energy security due to the decrease of conventional oil and gas resources in recent years. However, the complicated components such as wax, resins and asphaltenes in shale oil resulting in severe wall sticking and high pipeline back pressure during pipeline transportation. Therefore, it is necessary to study the influence of components on wall sticking of shale oil, which not only provide theory guidance for decreases the back pressure of shale oil pipeline, but also guarantees the safety of pipeline operation.

In this work, the effects of wax, resins and asphaltenes to wall sticking behaviors of shale oil under different temperature, stick-wall time and shear strength are discussed with the help of small scale stick-wall experimental device. To reveal cause of the change in the quality of the wall sticking, the composition and moisture content of sticky wall under different conditions are characterized. Furthermore, based on the theory of nucleation and adhesion, the influence mechanisms of components of shale oil on wall sticking are discussed systematically, so as to advance understanding for the shale oil transportation.

The results show that the lower of temperature and longer of sticking time, the higher of the wall sticking quality. The applying of shearing will reduce the quality of wall sticking quality. Meanwhile, it is revealed that the different contents in wax, resign and asphaltenes causing the difference in wall sticking quality under the same conditions. The synergy of nucleation of wax molecules and adhesion of asphaltene and resins, contributing to the wall sticking of pipeline. These conclusions provide theoretical guidance for ensuring the safety of the shale oil gathering and transportation pipeline in oilfield.

639284 Electrocoalescence of Water-Oil Mixtures Induced By Corona Discharge

Abstract:

Commercially available methods of oil/water separation are mostly energy inefficient, time consuming, and costly. This study considers a contactless method of electrocoalescence via corona discharge. Using a pin-to-plate configuration and pure/pulsed DC sources, non–uniform electric fields form which generate corona discharges. Applying the corona discharge to oil/water mixture results in a smooth separation of the two distinct phases. Considering the different processing parameters, it is possible to tune the effect of effective forces, *i.e.* electrophoresis or migratory coalescence (EP), dipole–dipole interaction (DDI), etc., acting on water droplets to facilitate the separation process. It is observed that the corona discharge significantly enhances the coalescence in concentrated W/O emulsions (made with silicone oil 350 cSt and 2 wt.% water content). After only 120 seconds of processing, it is observed that the two phases were separated into nearly two distinct phases. In the second part of the study, the effect of corona discharge on coalescence rate of water droplets in silicone oil 3000 cSt were investigated. Depending on the initial distance between the water droplets, coalescence could occur due to DDI for close distances or EP for far away droplets. For the case of DDI, it is shown that pulsed DC fields are preferred rather than pure DC. The coalescence times measured were 7, 3, 6, and 18 seconds for pure DC and pulsed DC with frequencies of 1, 10, and 20 Hz. For the case of EP, it was observed that pure DC performs better with coalescence times of 20, 40, 29, and 80 seconds for pure DC and pulsed DC in frequencies of 1, 10, and 20 Hz.

638956 1. Properties of H2 .2.H2 Safety 4. MOC Consideration for H2service.5. Hydrogen Embrittlement 6. High Temperature Hydrogen Attack.7. H2 Accidents 8. Lesson Learned.

Abstract:

HYDROGEN SAFETY:

1. Properties of hydrogen - introduction of hydrogen to the participants including but not limited to physical properties

2. Hydrogen Safety : element hydrogen is not easily catching fire as well its fire cannot be easily detected by naked eyes. Safety considerations will be presented.

3. MOC Consideration of H2 service- material when comes in contact with H2 can cause leak ofH2. the operating conditions and criteria will be presented.

5. Hydrogen embrittlement - industrial case studies will be discussed

6. High temperature hydrogen attack- pipe network portions impact due to hydrogen will be presented here.

7. H2 Accidents: Incidents during handling and using hydrogen will be shared.

8. Lesson Learned: learning during use of hydrogen will be elaborated to upcoming hydrogen projects.

639282 A Worker Safety Management System Based on Worker Location and Environmental Information during Tunnel Construction

Abstract:

Tunnel work is an environment in which 21 types of work are frequently performed, and it is a high-risk work at the construction site. A number of situations can occur, so thorough safety management is essential. However, due to the nature of the tunnel, communication stability is poor, so it is difficult to apply the existing IT safety management method to tunnel work. Inefficient safety management is being implemented. In addition, there are difficulties in emergency rescue in the event of an accident because there is no preparation for high-noise sites, so clear work instructions cannot be delivered, and the location of workers cannot be identified.

In this study, we developed a smart safety management system based on the method of positioning workers through wearable BIE module and wireless signal transmission/reception scanner and applied it to 15 sites for about 5 years to conduct real-time tunnel construction safety management and analyze its effectiveness.

Existing one-way communication system formed a vertical relationship between managers and workers, but this system enables two-way communication between workers and workers based on mobile, developing a horizontal relationship and greatly reducing safety accidents due to lack of communication. there was. In addition, by collecting environmental information (temperature, humidity, oxygen, carbon dioxide, carbon monoxide) in the tunnel in real time, it is possible to predict danger early, and it is possible to quickly deliver an evacuation command in case of a dangerous situation, and to rescue even at a tunnel construction site with a poor mobile communication environment. It can generate a signal so that the operator can protect himself. In addition, real-time risk assessment can be carried out by conducting a risk assessment according to the work situation and immediately reflecting the direct safety check result on the server. Awareness of safety and being able to act.

In addition to safety aspects, there were economic effects such as shortening the construction period. Compared to the existing method in which the safety manager visually identified and evacuated workers for blasting, one of the main tasks of tunnel construction, the real-time worker location monitoring method through the beacon attached to the worker's helmet takes about 20 minutes per blast. As a result, the construction period was shortened by about 30 days when calculated as 36 months of 2 blasting per day.

The introduction of the smart safety system in the tunnel made more efficient and automated safety management possible. The effect of planting water on 500 groups of 30-year-old larch per year is also expected by reducing manager's field trips and paper use.

640195 Fixing AI's Identity Crisis in the Industrial Space

Abstract:

Managing chemical processes at their optimal can often come undone when raw material variability, energy overages, recurrent and unplanned downtime, maintenance issues and ageing assets come into play. The sector is also under tremendous pressure to increase throughput while protecting their margins. Beyond their plant floor, the chemicals industry is also facing a sustainability challenge. According to the U.S. Energy Information Administration, industrial companies account for about 10% of global total final energy consumption and 7% of greenhouse gas emissions.

These challenges mean that the industry needs to optimize their batch processes in order to improve their top and bottom-line results. Artificial Intelligence (AI) and Machine Learning (ML) enable industrials to predict how to run their processes more efficiently to improve yield, quality and reduce downtime. As a result, AI is helping the industrials to reduce variation between batches and from plant to plant.

However, deploying AI is not without challenges. If not implemented properly, it can prevent the company from realizing the benefits of the deployment. Gartner says that 85% of the AI projects will continue to fail by 2022. Right from selecting the right AI use case to empowering the industrial workforce to use AI-powered systems and utilizing the right technology, the industry needs the benefit of best practices to achieve the benefits from AI.

This session will focus on three key best practices critical to taking the benefits of AI from aspiration to reality.

1. Understand your problems

One of the prerequisites of a successful AI implementation is identifying the operational challenges that keep you awake at night. How has the plant surprised you (good or bad) and do you know why? Are you satisfied with how your assets are performing? How well do you understand your data and why? By interrogating these questions helps you to identify use cases that will be both A) best suited for AI and 2) will deliver impactful value to your operations. One of the critical reasons why AI projects fail is because it is chosen to solve random problems. It is crucial to identify areas that will generate savings and revenue and have measurable Return on Investments (ROI). The use case should also be scalable in other functions or locations to increase the returns.

2. Understand the critical difference between AI and Industrial AI

Robust technology to manage scale and capability is required to ensure the success of AI initiatives. This is why it is critical to understand the difference between generic AI platforms and fit-for-purpose Industrial AI platforms. While generic AI platforms that are designed to be used by all types of businesses and industries (i.e. finance, healthcare, retail) and are built for common functions. Industrial companies need an Industrial AI platform that helps in resolving their specific problems. A generic AI platform is typically built to be used by data scientists and require coding skills. However, industrial companies are competing against a global shortage of data scientists and may not want to invest in hiring teams of them.

Industrials require an Industrial AI platform made especially for them because industrial data is complex with several variables and is collated from different data sources. Organizations generate around 3TB of data each day, but it is believed that approximately 40% of it is spurious and cannot be used. A generic AI platform is not designed to collate and work with data from so many different sources and the amount of noise that needs to be extracted from data. It is estimated that a single AI prediction can cost over $1000, which can become a costly and time-consuming exercise if your models feeding the predictions are built on noisy data and are not explainable. Instead, in order to achieve their desired impacts, industrial companies need an Industrial AI platform that cuts through noise and can extract clear visual trends to observe and action, prevent undesirable events before they happen, and go beyond model predictive control.

3. Invest in augmenting your workforce with AI

Possibly the most significant challenge faced by the industry in scaling AI is the lack of digital talent. There is an acute global shortage of data scientists, and it will be a few years before this problem is resolved. One solution is for industrials to hire consultants to execute AI deployments. But, even then, this approach may backfire because usually, the consultants are not domain experts and therefore can exhaust a lot of internal resource to contextualize the data so that it can be applied to the company's operational processes and assets. For AI to truly succeed and become ubiquitous across their operations, industrials need to augment their workforce with AI skills to reduce dependency on black-box tools and consultants. This is possible with today's no-code Industrial AI platforms, which automate the majority of data science functions, putting the operators in charge of their data and how best to extract valuable insights. The shortage of data scientists should not delay industrial companies from leveraging AI to realize their net-zero ambitions. This need is further heightened when the aging workforce comes into consideration, whereby institutionalizing knowledge and removing variability across plants becomes a priority.

This presentation will include real-world case studies of industrial companies applying AI to their processes including some batch operations. Examples that will be showcased include how a food manufacturer is using AI to accurately predict when the production of its animal feed additive product was completed in order to reduce batch-to-batch variability and maximize asset utilization.

By applying these three best practices, closing the gap in process variability is within reach with AI.

639937 Using Avert Physical Security (AVERT-PS) to Optimize Physical Security Effectiveness at Mission Critical Facilities

Abstract:

ARES Security Corporation (ARES) flagship product, AVERT, is an established suite of all-encompassing software solutions for the entire physical security life cycle from design to operations. AVERT transforms physical security planning and assessment from a dependence on the qualitative judgement of subject matter experts to a science based on a 3D digital twin, physics-based quantitative modeling, simulation, and artificial intelligence. The AVERT software suite is actively used by 65% of the USA commercial nuclear plants, along with industrial, corporate, and government organizations and has a proven track record of effectively assessing and optimizing security systems and identifying subsequent cost savings. One member of the AVERT software suite, AVERT Physical Security (AVERT-PS) is unique Security Risk Assessment (SRA) software used to perform a facility Vulnerability Assessment (VA) along with providing the capability to visualize, quantify, assess and optimize security posture. This solution's holistic and integrated approach, delivers accurate, measurable, and repeatable assessments of physical security design and operations. AVERT-PS provides security analysts with the capability to make security decisions based on quantitative, physics and probabilistic risk-based models and provide decision makers with a detailed understanding of the effectiveness of their physical security systems and operations against evolving threats. This paper will present a brief introduction to AVERT-PS along with example uses of the technology to review, improve, and optimize Physical Security Posture for mission critical facilities.

642995 Three Challenging Applications of Quantifying Critical Characteristics with Machine Vision (Poster)

Abstract:

In the era of Industry 4.0, Big Data, and Digitalization, manufacturers are increasingly seeking opportunities to gain actionable insight from their process data. Some key product or process characteristics however, remain difficult to measure with conventional sensors or rely on expensive or time-consuming sampling and lab measurement.

Machine vision solutions are a mature technology that enable reliable, online measurement of visual and thermal product properties. These systems are capable of quantifying critical product characteristics such as texture or size distribution that would otherwise be difficult to measure.

This presentation will focus on three challenging use cases and will provide an overview of the use case, hardware details, algorithm details, and results for each application. The applications include quantifying:

- *Texture of synthetic rubber crumb using color imaging*
- *Stirring intensity of hot metal using color videos*
- *Size distribution and moisture of iron ore using a polarization camera*

639323 Waste Plastics to Cleaner Fuels: Technical and Economic Assessment

Abstract:

An increase in energy demand in the recent decades has caused energy shortages. An increase in the fuel prices has forced the humanity to use the non-conventional energy sources for heat, power and clean fuel production. On the other hand, the waste generation has also increased with an increase in the population. Gasification and partial oxidation are among the few techniques that has a potential to convert the solid and liquid waste into syngas that can be transformed into hydrogen and other fisher tropsch fuels. Polyethylene (PE) and polypropylene (PP) are considered in this study because of their massive worldwide availability in the category of waste plastics. Simulation models are developed in Aspen Plus (R) to convert the PE and PP into syngas using the gasification process followed by the integration of gasification unit with the methane reforming unit to enhance the overall syngas production capacity. The technical and economic assessment is performed to convert the syngas into H2 and Methanol. This process integration approach not only enhances the overall H2 and methanol production rates but also reduce the reliance on the single fuel. It has been seen from results that the new integrated design containing both gasifier and reformer enhances the hydrogen mass production rate per mass feed rate by 7.5%. Furthermore, the process performance analysis showed that the efficiency of the new process is increased up to 2.5%, where the hydrogen production cost showed the reduction of 30% compared to the standalone gasification case. The results also showed that the new process has a potential of reducing the carbon dioxide emissions by 5.0%.

640072 Modeling Aqueous Solutions of Rare Earth Elements (REE) Salts with Association Electrolyte Nonrandom Two-Liquid Activity Coefficient Model

Abstract:

This paper presents an extension of the association electrolyte nonrandom two-liquid (AeNRTL) activity coefficient model [1] to calculate the activity coefficients for aqueous salt solutions of rare-earth elements (REE). Given the extensive consumption and limited supply of REEs in electronics, clean energy, aerospace, automotive and defense, the separation and recovery of REEs are of national importance [2]. Rigorous thermodynamic modeling for aqueous salt solutions of REEs is a key enabling technology in the research, development, design, and simulation of chemical processes for the separation and recovery of REEs.

Development of comprehensive thermodynamic models for the aqueous salt solutions of REEs has been very limited in the literature due to the lack of experimental thermodynamic data and the lack of rigorous electrolyte thermodynamic models capable of modeling ions of exceedingly high surface charge density including the REE ions. The recent advent of the AeNRTL activity coefficient model offers a major breakthrough in electrolyte thermodynamics especially for aqueous electrolyte solutions with high surface charge density ions [1]. This work reports compilation of the appropriate experimental thermodynamic data and accurate representation of these thermodynamic data for aqueous salt solutions of REEs and their mixtures with AeNRTL. The resulting model should be a valuable tool in support of the development of efficient and robust separation and recovery processes for REEs.

640209 Adsorption: A New Look at an Old Unit Operation

Abstract:

Adsorption: A New Look at an Old Unit Operation

While benzene is a widely used feedstock in the production of polymers, lubricants, dyes, pharmaceuticals, and pesticides; it is increasingly a cause for concern due to its carcinogenicity. Thus, removing trace amounts of benzene from contaminated chemical inventories becomes an important step in chemical processing. Such contamination may occur in plant storage tanks, transfer lines, and poorly flushed batch and semi-batch process equipment.

Removal of parts-per-million benzene through distillation is costly and often produces product and by-product streams that still contain measurable amounts of benzene. Similarly, liquid extraction could achieve the desired product purity but still involves an expensive distillation step to recover the solvent.

Using an adsorbent to remove benzene from a liquid solution can be both more effective and economical than distillation. However, benzene does not readily adsorb onto activated alumina, and activated carbon will non-selectively adsorb all hydrocarbons. Alternatively, molecular sieves can be used to adsorb benzene by taking advantage of their size exclusion properties.

In this presentation we show thermodynamic data and discuss the implications for the liquid-phase adsorption of benzene from 1-hexene and n-hexane on molecular sieves. Additionally, we discuss future work on the development of a continuous process in the form of a packed bed adsorber as well as modeling and scale-up possibilities.

AUTHOR INDEX

AIChE
120 Wall Street, FL 23
New York, NY 10005-4020

ISBN 978-1-7138-6833-0

CURRAN ASSOCIATES INC.
proceedings.com